MRSA and Staphylococcal Infections

MRSA and Staphylococcal Infections

Second Edition

by

Hernan R. Chang, M.D.

Also by Hernan R. Chang

Elysium: A Collection of Haiku and Senryu

MRSA- Spider Bites: The Flesh-Eating Bacterial Epidemic that Threatens America

MRSA and Staphylococcal Infections

 It is sold with the understanding that the author is not engaged in rendering medical, health, or any other kind of personal professional services in the book. Readers are advised to seek medical help when appropriate. It is the responsibility of a treating physician to determine medicine dosage and the best treatment for a given patient. The patient vignettes are the product of the author's imagination. Any resemblance to a real patient is purely coincidental. The naming of antibiotics, medicines, procedures, and tests in this book does not constitute endorsement for their use. Photographs were taken only after full consent was obtained from the patients. The author disclaims all responsibility for any liability, loss, or risk, personal or otherwise, which is incurred as a consequence, directly or indirectly, of the use and application of any of the contents of this book.

ISBN-13: 978-0-615-26274-1

Printed in the United States of America

Hope never abandons us,
Hope is stronger than fright

Contents

Foreword to the Second Edition

Since the publication of the first edition of this book, many patients, readers, friends, and colleagues have given me their suggestions. I have taken all those suggestions into consideration in writing this new edition of the book, and I hope this new edition is a better one. I have expanded several chapters, and I have added new ones.

Despite the large amount of information available through books, Internet sites, and libraries, more and more people with staphylococcal and MRSA infections present to their primary care physicians offices, urgent care centers, and emergency rooms (ERs). The disease seems to continue to spread.

Most people with a history of MRSA colonization get a reassuring answer when they ask about their contagiousness. The plain truth is that you are potentially contagious. Furthermore, being an MRSA carrier increases your risk of infection and death.

In order to understand the magnitude of this epidemic and to realize how the appearance of MRSA has and will influence our lives and our culture, consider the classic handshake. The handshake will never again be the same after MRSA.

Foreword to the First Edition

We are witnessing a worldwide increase in infections with *Staphylococcus aureus*—commonly called *S. aureus*, methicillin-sensitive *S. aureus* (MSSA), or simply "Staph." Many of these infections are due to invasive strains of methicillin-resistant *S. aureus* (MRSA). These infections cause significant morbidity and burden upon healthcare systems.

People not involved in a healthcare setting are often not aware of this silent epidemic. This lack of awareness includes the significance and danger of "community-acquired" MRSA infections and their potential threat to the healthcare system. The contents of this book are geared to the reader who does not have a medical or healthcare industry background. The information presented is intended to provide a general understanding of Staph infections. In order to curb the worldwide menace of Staph and MRSA infections, a significant behavioral change is needed. A well-informed public is essential for this behavioral change to occur. The opinions presented in this book are the author's and are not intended to replace the professional advice of a healthcare provider after a careful evaluation and clinical examination.

Chapter I

What Are Staph and MRSA?

What Is Staph?

Staph (also known as *Staphylococcus aureus* or *S. aureus*) is a bacterium that is commonly found in the environment and on the skin, nose, armpit, groin, and lower intestine of healthy people. This bacterium can cause illnesses ranging from minor skin infections (like pimples or boils) to life-threatening diseases such as Toxic Shock Syndrome. About one-third of the United States population carries Staph in their noses and may or may not have any symptoms or disease. These people are said to be "colonized" by Staph.

How Common Are Staph Infections?

Staph is one of the most common bacteria that produces skin infections in the United States and the world. These skin infections range from minor ailments, such as boils or other skin conditions, to severe cases such as pneumonia, bacteremia, Toxic Shock Syndrome, and "flesh-eating" disease (i.e., necrotizing fasciitis).

Why Do some People Become "Colonized" with Staph but Never Develop Symptoms or Disease?

People who are colonized with Staph have the organism living on or in the body, but do not have any signs or symptoms of illness or infection. They also may be colonized in multiple sites of the body but again with no signs of illness. Why is this? One reason may be due to an equilibrium that exists between the bacteria present in the nose, or other body parts, and the ability of the body to fight the bacteria. In addition, many other types of bacteria live on the skin along with Staph. These other bacteria take up space on the body and do not allow the Staph to completely take over. A strong immune system may also contribute to the prevention of Staph invasion to deeper areas of the body. It is important to remember, however, that regardless of whether a patient is colonized or infected with Staph, the organism can still be transmitted to another person, primarily through skin-to-skin and hand contact.

How Do People Who Are Colonized with Staph Develop a Staph Infection?

Anyone colonized with Staph can develop a Staph infection later on in his/her life. How? The bacteria are generally harmless unless they enter the body through a wound or cut, and even then, they usually only cause minor skin problems in healthy people. The bacteria can become deadly, however, when they enter the bodies of people who are already ill or have weakened immune systems.

What Is MRSA?

The common Staph is also called "methicillin-sensitive" *Staphylococcus aureus* or MSSA. MRSA stands for "methicillin-resistant" *Staphylococcus aureus*, which is a strain of Staph that

emerged in hospitals many decades ago. Typically, when a person gets a Staph infection, he/she is treated with antibiotics commonly used to treat skin infections. This new strain, however, was found to be resistant to an antibiotic called methicillin, which is why the new strain is called "methicillin-resistant" *Staphylococcus aureus* (MRSA). MRSA is also resistant to other antibiotics, such as oxacillin, nafcillin, penicillin, ampicillin, amoxicillin, and a group of antibiotics called cephalosporins. There are other antibiotics, however, that can be used to treat MRSA infections, such as vancomycin, linezolid, daptomycin, tigecycline, and quinupristin-dalfopristin. Although approximately one-third of the US population is colonized with Staph, only a minority of these are MRSA.

How Common Are MRSA Infections?

Back in the 1960s, MRSA was found in hospital settings among elderly patients and patients residing in nursing homes. However, in the last few years, a new type of MRSA (having the Panton-Valentine toxin) has emerged and affects people of all ages, many of whom have never been in a hospital. This new type of MRSA—originally found among wrestlers, football players, inmates, and military recruits—has now been found to be the most common cause of skin and soft-tissue infections among patients treated in ERs across the United States, according to a recent study (Moran GJ, et al. 2006. *N. Engl. J. Med.* 355:666-74).

How Frequent Are Colonizations with Staph and MRSA?

In the United States, approximately one-third of the population is colonized with Staph (89.4 million persons) and less than 1 percent is colonized with MRSA (2.3 million persons). Colonization rates with Staph are higher with the younger population, whereas colonization rates with MRSA are higher with the elderly.

How Can We Tell If Somebody Has Staph or MRSA?

Although people who have recurrent Staph skin infections tend to present with scabs in arms and legs, a swab from the skin, nose, or anal area, or a sample of body fluid (blood, urine, sputum, etc.), must be sent for culture to the laboratory to be able to say if somebody has Staph or MRSA. Based upon those results, the decision is made on how to treat the infection or colonization.

Can a Person Get Both Staph and MRSA Infections at the Same Time?

Yes, but this is extremely rare. Most people are colonized and/or infected with one type of Staph, but it is not impossible to have infection with both types of Staph at the same time in different areas of the body.

Chapter II

How Do We Get Staph and/or MRSA Infections?

Since approximately one-third of the population carry Staph bacteria somewhere on the skin, Staph infections are very common. Some studies have suggested that nasal colonization generally precedes the development of skin infections, and patients who have recurrent skin infections tend to be nasal carriers (Kluytmans J, et al. 1997. *Clin Microbiol Rev.* 10:505-20). Staph bacteria living on top of the skin can enter the body through a scratch, small wound, or a boil. This is probably the most common mechanism by which people present with Staph infections. The object causing the wound is introduced to the outside layer of the skin (where the Staph lives). As the object pushes deeper into our skin, it carries with it the Staph bacterium, which is then pushed deeper into the body tissues. Once it is deep inside, Staph can cause either minor or serious problems depending on a number of factors. Commonly, people who present with boils and are already colonized by Staph try to squeeze out the boil contents, and by doing so they, introduce the Staph in the skin, and the boil becomes a large abscess.

How Are Staph and MRSA Spread from One Person to Another?

Staph and MRSA bacteria are spread primarily through skin-to-skin contact with someone who has these organisms on his/her skin, but it can also be spread by touching objects that are contaminated with these bacteria. In other words, we touch a person or object that is contaminated with Staph or MRSA and then transfer this organism to our own body. How can this be prevented? Simply by washing your hands. Careful, scrupulous handwashing with soap and water, or waterless alcohol hand sanitizer, is the best defense against germs such as Staph and MRSA.

How Long Should You Wash?

If you are using soap and water, vigorously scrub your hands for fifteen to twenty seconds, rinse well, then dry with a disposable towel, and use the towel to turn off the faucet (see also Appendix A).

What Is the Proper Way to Use Waterless, Alcohol Hand Sanitizer?

Place a nickel-size amount of sanitizer in your hand. Rub the sanitizer thoroughly into your hands, paying particular attention to the thumbs and fingers, and then allow the product to air dry on your hands. Do *not* wipe the sanitizer off your hands with a towel or your clothing as air drying allows the product to kill the germs on your hands. The simple act of handwashing, either by soap-and-water method or using waterless, alcohol hand sanitizer, is the single most important thing you can do to protect yourself against disease and to prevent any possibility of infection after contact with potential sources of contamination.

MRSA and Staph infections have also been transmitted by athletic equipment (including equipment found in health clubs), tattoo equipment, razors, towels, and sheets that are shared by anyone who is colonized or infected with these organisms. Close-contact sports, such as wrestling and football, have also been contributing factors to the transmission of these organisms. Similarly, intimate contact with somebody who carries Staph in his/her skin will expose you to infection.

How Did Staph Become MRSA?

MRSA is a Staph bacterium that has become resistant to the antibiotic called methicillin. Actually, bacteria that have become resistant to antibiotics are on the rise throughout the world. A primary reason for this is the overuse of antibiotics. For years, physicians and hospitals have overprescribed antibiotics, using them unnecessarily and inappropriately. In addition, antibiotics are often prescribed for viral illnesses, which are caused by viruses not bacteria. Antibiotics cannot kill the viruses, and this, therefore, contributes to antibiotic resistance. Patients demanding antibiotics for themselves or their children every time they experience illness have also contributed to this overuse. Surprisingly, however, the vast majority of antibiotics produced in the United States aren't used for human consumption but instead go into animal feed for cattle, chickens, and pigs—not to treat illness of the animals but to accelerate growth and to prevent illness of those animals raised in overcrowded and unsanitary conditions. In addition, as antibiotics are excreted from these animals, the waste is often washed into streams, rivers, and groundwater tables.

Another way that bacteria become resistant to antibiotics is through their ability to mutate. Once bacteria are introduced to antibiotics, they can and often do become resistant to them.

Can Household Pets Transmit MRSA?

Yes, recent evidence has shown that household pets, such as cats and dogs, can transmit MRSA to their owners. The same might be true for Staph.

Can Food, Such as Meat, Transmit MSSA and MRSA?

Yes, Staph seems to be found in low amounts in meat sold to consumers. A recent study from the Netherlands showed that 2.5 percent of meat surveyed contained very low amounts of MRSA. Proper cooking probably gets rid of the problem and disease is unlikely. However, food handlers might be at risk for infection. In certain studies, high levels of Staph (not MRSA) have been found in raw chicken meat. In Japan, up to 65 percent of chicken meat was found to contain Staph (Kitai S et al. 2005. *J. Vet. Med. Sci.* 67:269-74).

Who Typically Gets Staph or MRSA Infections?

Infections with Staph and MRSA are more common and prevalent in hospitalized patients. These days, most patients admitted into hospitals are very sick, debilitated, immunocompromised, and even malnourished. These factors, coupled with the placement of devices such as indwelling, urinary catheters, IVs, and other medical and surgical procedures, increase the likelihood of a patient acquiring an infection. However, Staph and MRSA infections are also prevalent in dialysis centers, skilled nursing facilities, and long-term care facilities, where they manifest primarily as pneumonias, bloodstream infections, IV line infections, urinary tract infections, and wound and bone infections.

Can Healthy People Get Staph and MRSA Infections?

Yes, Staph and MRSA infections are also seen in otherwise healthy people who have either never been hospitalized or have not been hospitalized for a long time. These infections manifest themselves as skin infections with initial lesions resembling those of a spider bite (round, raised, reddened, and painful).

What Is "Community-associated" or "Community-acquired" MRSA (CA-MRSA)?

MRSA was never seen outside of hospitals or healthcare settings until 1999, when four previously healthy children who had never been hospitalized suddenly died of overwhelming MRSA infections.

People with MRSA infections are said to have "community-associated" or "community-acquired" MRSA infections (or simply CA-MRSA) if they meet the following criteria:

They have *not* had hospitalization, surgery, or dialysis or residence in a long-term care facility, skilled nursing facility, or hospice during the past year; they have *no* permanent indwelling catheters or percutaneous medical devices; they have *no* medical history of MRSA infection and colonization; and they have the diagnosis of MRSA made in the outpatient setting or by culture positive for MRSA within forty-eight hours after admission to a hospital.

How Common Are CA-MRSA Infections?

We have recently seen a marked increase in CA-MRSA infections in many parts of the United States and in the world. The reason for this is not entirely clear, but it seems that the bacterium is now more prevalent and also more virulent (some strains have certain toxins that make them more invasive). MRSA can also be transmitted in the healthcare setting (called healthcare-associated MRSA or HA-MRSA), and it happens primarily through the contaminated gloves or hands of healthcare workers or through contact among patients. Recent studies suggest that MRSA can also survive and replicate rapidly (thousand-fold) inside a common type of amoeba (*Acanthamoeba polyphaga*) that is present on most surfaces in the hospital environment. Amoebas can also spread in the air.

What Are the Risk Factors Associated with Staph and MRSA Infections?

Factors associated with an increase in Staph and MRSA infections include having poor hygiene, living in crowded places, having skin abrasions or cuts, or being in contact with infected items. Studies have shown outbreaks of CA-MRSA in children, military recruits, Native Americans, Alaskan natives, men who have sex with other men, and in people living in crowded environments and prisons.

Chapter III

What Diseases Can Staph and MRSA Produce?

How Is the Infection Produced?

Staph and MRSA infections are becoming more and more common in people who are otherwise healthy. Hygiene appears to play a role in whether or not somebody may get an invasive infection. The most common symptom of Staph or MRSA is the skin infection that starts as a boil or pimple. Many people blame it on a spider or insect bite. Interestingly, almost none of the people claiming such bites have seen the culprit. The common scenario is that they try to express the boil with their fingers. Sometimes pus is drained. But often it is not drained, and more inflammation is produced in the skin, leading to a furuncle or large boil, which subsequently gets larger and can produce a localized cellulitis (redness and pain in the skin). Depending on whether or not the bacterium is a very aggressive one (i.e., if the bacterium infecting the skin produces some toxins), impetigo, furuncles, cellulitis, darkening, blistering, or necrosis of the skin is seen. Most people do not wash their hands or the wounds at the same time they try to drain the pimples. If the Staph is already colonizing the skin, even those boils produced by bacteria other than Staph will end up being infected with Staph due to the poor hygiene.

Why Do Some People with a Staph Infection Get Very Sick?

After infecting the skin and depending on whether or not the person has a good immune response, the infection, after producing local symptoms, could proceed unchecked, enter the bloodstream, and produce bacteremia (growth of the bacteria in the bloodstream). Because the bacteria is not supposed to be in the bloodstream, certain cells in the body produce substances called "cytokines", which will produce fever and an inflammatory response. This inflammatory response is characterized clinically by fever, increased heart rate (tachycardia), increased respiratory rate (hyperventilation), and possible hypotension, which may herald the starting of what is called "septic shock." Some patients manifest with deadly forms of overwhelming infection or with the effects of Staph toxins, which are called Scalded Skin Syndrome and Toxic Shock Syndrome.

What Type of Diseases Can Staph and MRSA Produce?

The diseases that Staph and MRSA can produce are very similar, and they tend to be invasive. The diseases range from skin infections to pneumonia, infection of heart valves (endocarditis), infection of prosthetic devices (any device including pacemakers, hip or knee replacement devices, etc.), meningitis, spinal abscesses, and necrotizing fasciitis (necrosis of extensive areas of skin that can quickly become fatal). Staph and MRSA infections can also be seen after surgery or after indwelling devices are placed (including lines for IV antibiotics or parenteral nutrition, shunts placed in the brain for hydrocephalus, etc.). The mortality rate among patients with MRSA infections appears to be substantially higher than among those infected with Staph (MSSA).

Boil

A boil is an infection of hair follicles (also called furuncles). This infection leads to the accumulation of pus and necrotic tissue. If untreated, a boil could enlarge to produce an abscess. Treatment is local or with incision, drainage, and antibiotics.

Cellulitis

Cellulitis is an inflammation of the skin and subcutaneous tissues. It is mainly seen in the legs although any skin surface can be affected. Orbital cellulitis (around the eyes) and cellulitis of the elbow are also very common. The redness seen in the skin is not only related to the presence of infection but also to the release of toxins.

Impetigo

Impetigo is a common illness among children and is a superficial skin infection sometimes leading to the formation of a blister. Staph can produce bullous (blistering) and non-bullous impetigo. Treatment is with local antibacterials, such as mupirocin or if more serious, with oral or IV antibiotics.

Scalded Skin Syndrome

Staphylococcal scalded skin syndrome (SSSS) is also called Ritter's disease and Phemphigus neonatorum. It is caused by *S. aureus* exotoxins A and B. These exotoxins cause epidermolysis (detachment or loosening of the epidermis) and the formation of fluid-filled blisters on the body. Those blisters are very fragile and rupture easily. The blister is made of a very superficial skin layer formed after the detachment of the granulosum and spinosum layers. At the beginning of the disease, a positive Nikolsky sign can be seen (slippage of the superficial layer of the epithelium on pressure). Adults with SSSS have a poor prognosis. Children, on the other hand, have a good prognosis if treated adequately. Treatment is with supportive care, hydration, prevention of skin superinfection, and antibiotics.

Toxic Shock Syndrome

Toxic Shock Syndrome is caused by *S. aureus* exotoxins and other bacterial products acting as superantigens. These superantigens cross-link T-cell receptors with class II major histocompatibility complex, leading to an activation of large numbers of T-cells. This activation results in the release of large amounts of cytokines, producing hypotension, fever, nausea, vomiting, diarrhea, malaise, peeling of the skin, and multiorgan system failure. Death is commonly seen with this syndrome. Treatment is with antibiotics (clindamycin is commonly included), supportive measures, and sometimes IV immune globulin.

Pneumonia

MRSA and MSSA can produce a severe, necrotizing pneumonia that quickly leads to respiratory failure. It is seen more and more frequently. Treatment is with antibiotics that penetrate the lung. Some people advocate linezolid for the treatment of MRSA because of its lung penetration, but it is not bactericidal. Daptomycin should not be used to treat MSSA or MRSA pneumonia because it does not penetrate well the lung.

Endocarditis

Endocarditis is an inflammation of the endocardium—the inner layer of the heart. It can be seen in a natural or prosthetic valve or in other areas of the heart cavities. The result of this infection is the formation of a vegetation composed of fibrin, platelets, and bacteria. It can be from a few millimeters to several centimeters in length. Treatment sometimes requires surgery to excise the vegetation or to replace a damaged heart valve. Long-term antibiotics (at least six weeks or more) are needed. Prognosis is dismal in elderly persons. Intravenous drug abusers are at risk for endocarditis, and often they present with endocarditis of the tricuspid valve.

Infections of Prosthetic Devices

These days, an increasing number of infections are seen in prosthetic knees, shoulders, and hips. Commonly, the patient undergoes the surgery without problems and one or two weeks later presents with fever and tenderness and redness (cellulitis) in the surgical area. Treatment is with incision and drainage of the concerned area if the infection is superficial or if the prosthetic device is infected, with its removal and long-term, IV antibiotics. Also, patients who have indwelling catheters (such as hemodialysis lines) are at risk for staphylococcal infections.

Meningitis

Meningitis is inflammation and/or infection of the meninges—the membrane that covers the brain and spinal cord. The cerebrospinal fluid is enclosed by this membrane. Diagnosis is made by performing a lumbar puncture (spinal tap) and by sampling and studying the cerebrospinal fluid. Treatment is with antibiotics that are able to enter and concentrate effectively into the central nervous system.

Necrotizing Fasciitis

Necrotizing fasciitis is infection of the fascia in subcutaneous tissues and results in necrosis (death) of the involved tissues. Muscles and tendons are also commonly affected. The patient has malaise, pain, fever, and leukocytosis (increase in the number of leukocytes in the circulating blood). Bacteria other than *S. aureus* can produce necrotizing fasciitis, including streptococci, *Clostridium,* and *Vibrio*. Treatment entails surgical resection of the necrotic tissues and IV antibiotics.

Is MRSA More Aggressive Than Staph?

Yes, CA-MRSA is more dangerous and virulent than Staph (MSSA) and is more difficult to treat. The current epidemic strains of CA-MRSA have the ability to produce certain toxins that can deactivate white blood cells (i.e., Panton-Valentine leukocidin toxin), leading to severe skin disease (necrotizing fasciitis) or necrotizing pneumonia. This toxin is less often found in HA-MRSA. CA-MRSA tends to be more susceptible than HA-MRSA to antibiotics, with the exceptions of beta-lactams and erythromycin.

Chapter IV

How Do We Diagnose a Staph or MRSA Infection?

Staph or MRSA infections are revealed by collecting samples in a sterile environment and sending them to the microbiology laboratory for culture. MRSA is revealed by showing that the Staph in culture is resistant to oxacillin in the petri dish. To do that, the bacterium needs to be cultured from the sample. Once it is isolated, it is exposed to the antibiotics to determine if there is resistance to any of them. Some new molecular tests are used to speed up the diagnosis. They can be done in a few hours but are much more expensive. They use DNA detection-based, polymerase chain reaction techniques. The sample should be taken aseptically from skin lesions, blood, urine, or sputum. It can also be taken from sterile body fluids (cerebrospinal fluid, pleural fluid, or joint fluid). In general, any body tissue or fluid can be cultured. Lines and indwelling devices can also be cultured. Additionally, after surgery, if a prosthetic material is thought to be infected, it can be sent for culture.

How Do We Diagnose Staph or MRSA Colonization?

Colonization with Staph or MRSA is typically diagnosed by wiping both nostrils with a swab and sending the swabs to the laboratory for culture. A sterile, moistened swab is rotated in each nostril two to five times clockwise and counterclockwise. The swab should go up to three-fourths of an inch into the nostrils. The idea is to harvest squamous epithelial cells inside the nose. Many people have the tendency to touch their nose, and children do it very frequently. Since the Staph or MRSA inhabits the skin, the fingers carry bacteria, which subsequently remain in the moist environment of the nose. Swabs for cultures can also be taken from other sites, such as the armpits, groin, or anal area. Recent studies suggest that the throat may also serve as a reservoir for MRSA.

How Do We Decolonize a Person?

If the nose is colonized, an ointment (mupirocin) is called for. Additionally, oral antibiotics are taken for a period of at least seven days. Subsequently, cultures are repeated to determine whether or not the bacterium has been eradicated. This eradication may be transient since reinfection can occur as a result of manipulation of the nose with infected/colonized fingers. Infections with S. *aureus* and MRSA are more frequent in people who are already colonized. Therefore, it is important to try to reduce the risk of infection by pursuing decolonization.

Chapter V

How Do We Treat Staph and MRSA Infections?

Staph and MRSA infections are treated with antibiotics. If there is any collection of purulent material (pus) in the skin or in an internal organ or cavity, it needs to be drained.

Antibiotics are not able to dissolve the pus that already exists in a large abscess. Therefore, such an abscess needs to be surgically drained or drained with a needle and the help of radiographic techniques, such as computer tomography or ultrasound. In the case of infections that are not severe, a short course of antibiotics may suffice. However, when infections are deep or severe, antibiotics, preferably delivered intravenously, are needed for several weeks.

Infections with Staph and MRSA need to be taken seriously as they have the potential to be lethal within a short period of time if not treated promptly. The strains responsible for CA-MRSA infections seem to have an increased virulence, and in matter of days, infections with CA-MRSA can produce large abscesses.

Treating MRSA

Among the antibiotics useful in the treatment of MRSA, vancomycin, an IV antibiotic, has traditionally been used. Resistance to vancomycin is extremely rare, and only a few cases have been reported. Other antibiotics that can be used to treat MRSA include linezolid, daptomycin, tigecycline, clindamycin, trimethoprim-sulfamethoxazole, doxycycline, rifampin (not recommended alone), and quinupristin-dalfopristin. Some of these antibiotics are used "off-label," meaning that they have not been approved by the FDA for treating MRSA infections specifically.

Vancomycin

Vancomycin has been used for decades to treat MRSA infections. It is available in IV form (the IV preparation can be used orally to treat diarrhea associated with *Clostridium difficile* infection), and its levels need to be monitored to prevent toxicity and to achieve therapeutic levels. Very little resistance has been observed to this antibiotic. It can produce "red-man syndrome" when given intravenously, which is due to the sudden release of histamine by mast cells. This can be minimized by giving the infusion slowly over several hours and using antihistamines. Some people develop toxicity when levels are very high or as an idiosyncratic reaction, but these reactions are uncommon when levels are monitored carefully and doses are adjusted accordingly.

Linezolid

Linezolid is a relatively new antibiotic (an oxazolidinone) that comes in oral and IV versions making it attractive for treatment in an outpatient setting or when the time comes to discharge the patient after hospitalization. It is an expensive antibiotic. It has effective penetration in tissues and is touted as a good choice to treat MRSA skin infections and pneumonia. It seems to have better penetration than vancomycin into the bronchoalveolar tissue of the lungs. However, it is bacteriostatic (inhibits the growth of bacteria without destruction) not bactericidal (able to destroy bacteria). Linezolid also has the potential ability to decrease the production of Panton-Valentine leukocidin toxin (clindamycin is also able to do this). It can be used safely in patients with renal failure, when patients are allergic to vancomycin, or when outpatient or oral therapy is desired. A patient receiving linezolid needs close monitoring as it can produce myelosuppression (anemia and a decrease in the level of platelets, which is reversible within a few days after stopping the medication). It can also produce serotonin syndrome, lactic acidosis (by inhibition of mitochondrial protein synthesis), toxic optic neuropathy, and peripheral neuropathy.

Daptomycin

Daptomycin is a cyclic lipopeptide antibiotic. It is also a relatively new antibiotic that can only be used intravenously. It is bactericidal against MSSA and MRSA. Daptomycin resolves bacteremias faster than vancomycin and other antibiotics. Its dose has to be adjusted for renal insufficiency. It can produce muscle damage and elevation of liver enzymes. Weekly monitoring of muscle enzymes and liver enzymes is recommended. This too is an expensive antibiotic.

Tigecycline

This new antibiotic—a derivative of minocycline—is effective against MSSA and MRSA and other bacteria. Hence, it can be used on polymicrobial infections. Its main side effects are nausea and vomiting.

Clindamycin

Clindamycin is an old antibiotic that affects protein synthesis in bacteria. It has been said that it can inhibit toxin production. It has good penetration in tissues, including bone. However, in some Staph and MRSA strains, its resistance can be inducible, meaning the bacterium will become resistant during treatment. Inducible resistance is commonly tested by using the D-zone test. A common side effect is diarrhea. In fact, the first cases of *C. difficile* diarrhea and colitis were reported on patients treated with clindamycin.

Trimethoprim-sulfamethoxazole

This is an old antibiotic in which two antibiotics act sequentially in a metabolism pathway to produce a synergistic effect on susceptible bacteria. The antibiotic is bactericidal against susceptible strains of Staph and MRSA. It can produce renal failure and hyperkalemia (high concentration of potassium in the blood), so renal function and potassium levels have to be monitored while on treatment. Oral treatment (two tablets per day) may not be adequate to treat infections with large bacterial burden. Trimethoprim-sulfamethoxazole has been used with rifampin to treat colonization.

Doxycycline and Minocycline

Doxycycline and minocycline belong to the tetracycline family and have been around for a long time. Availability of minocycline is difficult these days as production was halted recently. These antibiotics are bacteriostatic and are commonly used with rifampin to treat colonization. Common side effects include tinnitus and dizziness. When using them, precautions need to be taken to avoid sunburn as the antibiotics produce photosensitivity. They can also produce severe irritation in the esophagus. Each capsule should be taken with a full glass of water, and the patient should not lie down for thirty minutes after taking these medicines. They should be taken with food if stomach upset occurs. These medications should be taken two to three hours before or after taking any medications containing magnesium, aluminum, or calcium. If symptoms persist, the patient should stop taking them and call his/her healthcare provider.

Rifampin

Rifampin is an old antibiotic that traditionally has been used to treat tuberculosis. Many strains of MSSA and MRSA are susceptible to this antibiotic. Unfortunately, resistance can develop during treatment. Therefore, it is not recommended to use alone or when the bacterial load is high so that the chance of resistance is lower. Rifampin can produce discoloration of urine and contact lenses. It is used in combination with trimethoprim-sulfamethoxazole or doxycycline to treat colonization.

Quinupristin-dalfopristin

This antibiotic is a streptogramin (an antibiotic produced by a bacterium of the genus *Streptomyces*) and has bactericidal activity against several bacteria, including MSSA, MRSA, and vancomycin-resistant *Enterococcus faecium* (VRE). It is used only intravenously and preferably through a central vein catheter as it can produce thrombophlebitis when used through a peripheral vein access. Its use can lead to severe myalgias (muscle pain) and arthralgias (joint pain). It can also interfere with the metabolism of several drugs by inhibiting the cytochrome P450 (3A4). This medication needs to be adjusted in cases of liver insufficiency.

Ciprofloxacin and Levofloxacin

Ciprofloxacin and levofloxacin are fluoroquinolones (antibiotics that inhibit the replication of bacterial DNA) that have been around for quite some time. Some strains of Staph remain susceptible to these medications. Ciprofloxacin can be obtained for free with a prescription in some pharmacies. They can produce diarrhea and, in some people with low magnesium levels, rupture of the Achilles tendon. If a patient experiences muscle pain, cramps, or diarrhea, he/she should stop taking the medications and call his/her healthcare provider. Antacids can inhibit their absorption, so they should be taken at least three hours before or after antacids.

Gentamicin

Gentamicin is an aminoglycoside (a group of antibiotics that inhibit bacterial protein synthesis) that is commonly used in the treatment MSSA and MRSA infections, in combination with nafcillin or vancomycin to help "clear" the bacteremia. There is a scarcity of data on its effectiveness in combination, but many doctors have used it with success. Gentamicin must be used cautiously in patients with renal insufficiency or with hearing problems,

Newer Antibiotics

Newer antibiotics include dalbavancin, televancin, oritavancin, and ceftobiprole. Some are still in the investigational stage.

Treating MSSA

The following antibiotics are used to treat MSSA. They cannot be used to treat MRSA successfully though most antibiotics used to treat MRSA can be used against MSSA.

Nafcillin and Oxacillin

Nafcillin and oxacillin are available only in IV form. They are narrow-spectrum, beta-lactam antibiotics of the penicillin class active against MSSA. MSSA is exquisitely sensitive to these antibiotics. In endocarditis, their use is preferred to other antibiotics because of the high sensitivity of MSSA to them. They need to be given frequently (every four hours) due to their short half-lives. Side effects include diarrhea, thrombocytopenia, or leukopenia. They are sometimes used with an aminoglycoside (i.e., gentamicin) for synergy.

Cefazolin

Cefazolin is also available only in IV form. It is a first-generation, cephalosporin antibiotic. It is used every eight hours and usually produces fewer side effects than nafcillin.

Dicloxacillin and Flucloxacillin

Dicloxacillin and flucloxacillin are oral antibiotics used four times a day. Like nafcillin and oxacillin, they are narrow-spectrum, beta-lactam antibiotics of the penicillin class. Diarrhea is a common side effect.

Cephalexin

Cephalexin is an oral antibiotic also used four times a day. It is a first-generation cephalosporin. Diarrhea is also a common side effect.

Chapter VI

How Do We Prevent Staph and MRSA Infections?

The cornerstone for prevention of infection with Staph and MRSA is the implementation of the simple and straightforward measures indicated below:

You should wash your hands with soap and water regularly, particularly after using the toilet and before meals. You can use an antiseptic soap, but there is not much data to support the idea that it is better than plain soap and water. Use a hand sanitizer with alcohol if soap and water are not available. The goal is to maintain clean hands. Good hygiene and daily bathing should be maintained.

Clean cuts or scrapes that bleed immediately, and cover them with a bandage until the skin is intact again. If redness appears or pain increases, seek medical help.

Do not touch other people's broken skin, boils, or infected or bandaged areas. If you must, wash your hands before and after touching their skin.

Do not share towels, clothing, bar soaps, shampoos, nail clippers, cosmetics, razors, or toothbrushes.

Do not share sports or exercise equipment that has not been cleaned between users. Shower immediately after working out or competing in contact sports.

Discourage allowing pets to share beds with members of the household as there is evidence that household pets might transmit MRSA. If your pets play in the dirt or are outdoors frequently, sharing beds or sofas with them is probably not a good idea. If you do share, make sure your pets are bathed properly.

Try to avoid walking inside or outside in your bare feet, and if you do so, wash them regularly.

Do not ignore an insect bite, spider bite, or a persistent sore anywhere on the body. See a healthcare provider immediately if severe itching or pain develops. Don't wait for blistering before seeking treatment.

If You Have a History of Being Colonized with MRSA or Have Had an Infection with MRSA, Are You Contagious?

The plain truth is that you are potentially contagious. You can disseminate it to anything that you touch (objects and people) if your hands are not regularly cleaned/washed (remember, bacteria is on your skin). If you cleanse your skin regularly, bacteria cannot multiply, and you cannot transmit them. Therefore, there is no need to panic. However, being a MRSA carrier increases your risk of infection and death. Yes, death. MRSA is a very virulent and invasive bacterium that if not acted upon promptly can kill pretty quickly. We can't, however, predict who will have a rapidly progressing infection and who will have a more indolent infection.

If you are already colonized, follow the instructions of your healthcare provider, which may include applying a mupirocin ointment to the inside of your nose, taking oral antibiotics, and practicing good hygiene. Additionally, use shampoos and antiseptic soaps containing chlorhexidine. Some people discourage the use of chlorhexidine as a shampoo or in use on the face and ears as some cases of toxicity have been reported. For specific measures, follow the advice of your healthcare provider. These measures should help prevent the number of recurrences of infection. Note that resistance of MRSA to mupirocin has been documented. You should use oral antimicrobials in conjunction with the mupirocin ointment to help in the eradication of MRSA from the nostrils. Combinations of rifampin and trimethoprim-sulfamethoxazole or rifampin and doxycycline are also commonly used.

If you have a history of recurrent infections with Staph or MRSA and have a sudden, unexplained onset of chills or high fever, contact your healthcare provider as this might indicate an underlying infection.

What Do You Do When You Go Home If You Are Colonized or Have Been Infected with MRSA?

Follow the guidelines regarding hygiene. Your family and household members, if they are not immunocompromised and if they do not have a chronic illness and/or skin disease, are unlikely to get infected if they respect hygiene measures. You should not share towels, toothbrushes, razors, clothes, or shoes. Family and other household members should probably be tested to see if they are colonized or not. Anyone who is colonized should seek decolonization.

If You Are Colonized or Have Been Infected with MRSA and Then Go Back to the Hospital, What Do You Do?

Tell the hospital that you have been diagnosed with either colonization or infection with MRSA so they can test you again and place you on Contact Precautions to prevent the spread of the bacterium.

How Can We Prevent Transmission of Staph and MRSA in the Healthcare Setting?

In the healthcare setting, prevent transmission from patient to patient through the use of Standard Precautions and Contact Precautions. The goal is to prevent infection by direct contact and droplet spread.

The patient with MRSA infection (and MRSA colonization) is placed on Contact Precautions, preferably in a private room, or, if not possible, in a room with another patient with MRSA but no other infection. A patient is placed on Contact Precautions before demonstration by cultures if there is even a hint or suspicion of MRSA infection. Staff and visitors wear gloves and gowns when entering the room, and wear gloves when handling all devices and laundry. They use masks if the patient has pneumonia or if there is a possibility of transmission by droplets (such as changing dressings in abdominal wounds or if the patient is sneezing). Hand washing or using sanitizer is mandatory before entering and after leaving the room. Stethoscopes and other materials in their rooms should not be shared with other patients. Dishes and dietary trays are not a major source of disease transmission.

Dispose of contaminated dressings and devices properly in specific containers. Clean contaminated equipment and surfaces with a solution of one part bleach to ten parts water or a commercial disinfectant. Wash linens in hot water and dry with a high-temperature dryer setting.

Is Surveillance Culturing and Placement of Patients in Contact Precautions Useful?

Active surveillance culturing to identify MRSA-colonized patients and the use of Contact Precautions have been advocated and used in the healthcare setting successfully. These measures should help to reduce the spread of the multidrug-resistant strains of MRSA seen in hospitals. Although it is well known that antibiotic usage exerts pressure to select for resistant flora, patients acquire MRSA in the hospitals primarily due to contamination. This underscores the importance of contact precautions and active surveillance cultures.

Who Needs Screening for Colonization in the Healthcare Setting?

Any patient who meets one of these criteria is placed on Contact Precautions until screening culture results are obtained. This is not a complete list as criteria can vary from hospital to hospital:

undergoing dialysis
undergoing joint replacement procedures
undergoing open spinal procedures
undergoing cardiovascular surgery involving a mediastinal operative approach.
admitted from a nursing home or assisted-living facility
admitted from a homeless shelter
admitted or transferred into ICU/CCU
admitted from a rehabilitation hospital
admitted from jail/prison
aeonates transferred into a NICU from an outside hospital
history of HIV infection
diagnosis of skin or soft-tissue infection at admission
history of IV drug abuse
previous history of MRSA infection or colonization

Will Colonization with MRSA Ever Go Away? Am I Potentially Contagious Forever?

Colonization with MRSA sometimes goes away with a combination of good hygiene and other factors. If subsequent tests for colonization are negative, Contact Precautions will be lifted. But remember, you can be infected and/or colonized again! Some studies suggest that in most cases, colonization is reestablished within six to twelve months, so the tests for colonization might need to be repeated and decolonization might need to be done again. Also, it is possible that the colonization test was not able to detect colonization in a specific site (nostrils), but in fact you are still colonized in other areas of the body.

Chapter VII

Dealing with MRSA at Home

If you are colonized and/or infected with MRSA, here are some suggestions to reduce the possibility of transmitting it to your family and friends:

Clean your house using a commercial disinfectant or a dilution of one part bleach to ten parts water. The bleach solution needs to be prepared fresh daily. Disinfectants kill MRSA effectively provided that enough time is given for the solution to act on the bacterium. Let the surfaces where the disinfectant was spread air dry. Use gloves while doing this, and wash your hands with antibacterial soap after taking off the gloves. Wash your hands for at least fifteen seconds.

Use gloves to handle body fluids, stools, or soiled clothes. Sinks, toilets, and tubs have to be cleaned daily. If possible, the person with MRSA should have a personal bathroom.

Wash your hands before preparing food, before eating and drinking, before and after using the toilet, after smoking, after coughing, after sneezing, or after touching anything that a MRSA patient has previously touched.

When eating, do not share dishes or utensils. When washing them, use soap and hot water or a dishwasher.

Chapter VIII

Dealing with MRSA Outside the Home

If you are going to have a haircut, insist the combs, scissors, and razors are disinfected prior to use on you. If razors are used to shave hairs on your neck, they need to be disinfected so you don't get the microbes from the last customer's skin.

If you are going to have a pedicure or manicure, all instruments that will come in contact with your body need to be sterilized or new.

If you go to a restaurant, utensils need to be clean. Don't share your food with others. Wash your hands before eating. (How many people have you seen wash their hands before they go to eat in a restaurant?)

If you go to a public restroom, wash your hands after using the toilets or urinals. (How many people have you seen *not* wash their hands after using the restroom in, for instance, a movie theater?)

If you have used a public swimming pool and want to take a shower later, do not go barefoot in the shower. Use sandals and do not share soaps or towels with anybody.

If you use a public facility or a device that is used by other persons, such as a computer keyboard, wash your hands before leaving that place. Alternatively, you can use a hand sanitizer. In the years to come, carrying an alcohol-based sanitizer will probably be as common as carrying lipstick or tissues.

The classic handshake greeting is a part of our culture. However, that might not be a safe practice these days. We never know what people have done before shaking our hands and/or if they wash their hands systematically or not. Also, intimate contact (as during intercourse) and close contact (as in sports) can definitely allow for transmission of Staph and MRSA.

If you are a patient in a hospital, insist that anyone seeing you, including your doctors, use a sanitizer before examining you or before putting gloves on their hands. They should also wash their hands after removing their gloves.

Chapter IX

Case Studies

These passages are meant to convey typical scenarios, symptoms, and treatments of Staph and MRSA infections.

Patient Vignette 1

Mark is a thirty-five-year-old single male who lives in the suburbs of Chicago. He works in a plastic factory where he handles boxes loaded with the finished products, which are exported or sent to other locations in the United States. He uses gloves at work and does not smoke, drink, or use any drugs. He has a dog that is three years old. His dog sleeps with him, and Mark has noticed some lesions on his dog's skin. He is aware that his dog has skin allergies and sometimes has severe itching due to flea allergies. He uses medication to treat the dog's condition every three months.

Mark has noticed lately that he has small lesions on his arms and shins, which appear to come from nowhere, and he develops localized swelling whenever he scratches them. He thought for some time that these were allergic reactions to his dog's dander and increased the bathing of his dog to once a week.

However, he still sees the small lesions from time to time. They improve after a few days, but there are always small scabs that form one or two days after he scratches them. He has noticed that some of them itch and appear to be small blisters similar to spider bites that subsequently enlarge. However, he has not seen any spiders in his house or bed.

It was a Friday afternoon, and Mark wanted to go out with friends to a local restaurant for dinner. He noticed that he had a small pimple on his right leg that had appeared that morning. He had carefully washed it out and had applied some over-the-counter antibacterial ointment to it. He noticed that there was a small area of swelling around the lesion. He also experienced some increased itching. He took a shower and went to dinner. He had a good time, drank two beers, and went home.

The next morning, he awoke with extreme pain in his right leg and felt feverish. He looked at his leg and saw that there was a large, red, swollen area. He touched his leg and felt severe pain. He changed his clothes and went to the hospital's ER. He was evaluated by a physician and had some blood work done. Then the doctor informed him that he needed to remain in the hospital for IV antibiotics because he had a high fever, and his blood work showed an increase in white blood cells. A technician took blood samples and blood cultures from him. He was asked if he was allergic to any antibiotics. Since he had no allergies to any antibiotics, he was given an IV antibiotic called cefazolin. A nurse did cultures of his wound, which was darker and draining pus. His leg did not seem to improve with the antibiotic and was getting more swollen. Thereafter, another doctor told him that his antibiotic would be changed to a more powerful one called vancomycin. The pain in his leg was not reduced with the painkillers, but nevertheless, he tried to sleep. He had an MRI done on his leg to see if the infection had rapidly progressed and affected his bones. The next day, the redness and swelling in his leg were a little better. The pain, however, was still there and was severe. That afternoon, he was told by hospital staff that people coming into his room would need to wear gloves and gowns. He got worried, thinking that he might have some highly contagious disease. He was told that it was the hospital's policy, that the infection was rather commonly seen, and that its name was MRSA.

The doctor that saw him the day before told him he needed to culture his nose to see if the MRSA was hiding in his nostrils. He took swabs of the inside of Mark's nostrils. During the procedure, Mark was trying hard not to sneeze and started to have profuse tearing. With the new antibiotic, his leg was rapidly improving and the swelling was going down. The MRI of his leg came back negative for bone infection, and he was told that he needed either IV or oral antibiotics at home for at least seven more days.

The next day, Mark's leg was almost back to normal except for the dark scab and the ulcer. He was told that his nose culture was positive for MRSA, and he needed to apply an ointment to his nostrils with a cotton tip three times a day for at least seven days. The name of the ointment was mupirocin, and he was also told that he needed to take two pills a day for seven days. Their names were rifampin and trimethoprim-sulfamethoxazole. One of them would make his urine dark orange in color. His blood cultures came back negative, and no further workup was planned. He was getting anxious as he needed to go back to work. Since his leg was looking much better, he was finally discharged and told to take another pill called linezolid for seven more days and to follow up with his primary care physician. He was also told to try to shower and shampoo his hair with a chlorhexidine gluconate solution.

Mark's case is a very common one these days. It is just one of the hundreds of cases of CA-MRSA infection seen daily in ERs. Mark sought prompt care, was treated adequately, and recovered successfully.

Patient Vignette 2

Jeanne is seventy years old and lives in New York in an assisted-living community. She has diabetes mellitus and heart disease. She had a coronary bypass graft six years ago. She has also had two knee replacements. She had a right hip replacement due to a fracture following a fall the year before her heart surgery. She still has some problems ambulating and has occasional pain in her hip. Her blood sugar level has been well controlled and is in the 105 to 110 range in the mornings. She has no kidney disease or diabetic neuropathy although she suffers from glaucoma and has had vision troubles.

It is late December, and she received flu and pneumonia shots two months earlier. She has been doing well, despite the fact that some of the people in her assisted-living community were complaining of colds, and two residents have recently had pneumonia for which they needed to be hospitalized.

Jeanne finished lunch and went to take a nap in her room. She awoke about three hours later with chills and nausea. She also felt dizzy, so she called the personnel in her assisted-living facility. An aide came to her room and took her temperature. She had a fever of 102 degrees. After calling her doctor, the personnel decided to send her to the ER of the nearby hospital. She agreed. She was feeling more and more ill. She was short of breath and beginning to cough.

After she was transported to the ER, she was evaluated by a nurse and then by one of the physicians on duty. She then had a temperature of 104 degrees. Her white blood cell count was elevated, and her chest X-ray showed a right-side pneumonia. She was then admitted to the hospital. Cultures of her blood and urine were taken.

Jeanne was beginning to cough more and more. She was short of breath and had some chest tightness. She had an electrocardiogram that showed some abnormality, suggesting the possibility of a heart attack. Her oxygenation was also poor, which led to her having oxygen delivered by a mask. She was soon wheezing constantly. She was given a diuretic and some bronchodilator nebulizations. After a few minutes, she began to feel better. She also received two IV antibiotics.

Because of the various medical problems she was facing, she was admitted to the Intensive Care Unit (ICU). Her breathing was still labored two hours later. A cardiologist on call saw her and recommended more diuretics, as well as an infectious diseases consultation, following his observation that she had a high fever, and her blood pressure was dropping.

Her low blood pressure prompted the use of medications to increase it. Also, a catheter was placed by the cardiologist in her neck to better evaluate her cardiac condition. After a few minutes, the cardiologist concluded that Jeanne was experiencing an early case of sepsis.

The infectious diseases specialist arrived one hour later and recommended that more antibiotics be started, requested a CT scan of Jeanne's chest, and requested that an echocardiogram be done in the morning. Jeanne was feeling increasingly weak, and her breathing remained heavy despite all the treatments.

Three hours later, she was on an oxygen mask and still unable to breathe properly. Additionally, her urine was starting to diminish (she had a urine catheter placed earlier to measure her urine). The decision was then made by the attending physician to put her on a mechanical respirator, and she agreed. She was fully sedated before being placed on the mechanical respirator.

The next morning, her sedation was reduced slightly to see if she was able to understand and follow commands. Her CT scan showed she had pneumonia in her right lung. Her echocardiogram did not show any sign of infection in the heart valves, but her blood cultures came back positive for possible Staph infection. Twenty-four hours more were needed to have the cultures finalized and to ascertain whether or not she had a Staph infection. Her blood pressure remained low, and she was on three medications in an effort to increase it. Her urine output had become very minimal, and a consultation was requested with the renal consultant (nephrologist).

Jeanne was on the ventilator, and samples of her lung secretions (sputum) were sent for culture to see what type of bacteria may have caused her pneumonia. Her urine cultures were negative for bacteria, so she was started on tube feedings using a tube going from her mouth to her stomach. The culture of her sputum demonstrated the presence of MRSA.

Over the next few days, her clinical condition did not change much. Her blood cultures also turned out to be positive for MRSA. She continued on the IV antibiotics, and subsequent blood cultures were negative, suggesting that the antibiotics were effective against MRSA. However, she continued to retain fluid and had no urine output. The renal consultant recommended dialysis until her kidneys recovered. She agreed to the placement of a catheter in her left groin to enable the dialysis. The first session of dialysis proceeded well, and her apprehension was reduced. She was still intubated and anxious since she could not talk. Her frustration came from the fact that she could not communicate much except to nod yes or no when the nurses and doctors asked her questions.

One week later and after three sessions of dialysis, her clinical condition clearly started to improve. She began to have a scant amount of urine, signaling that her kidneys were improving. Her chest X-rays, however, still showed pneumonia, and some fluid started to accumulate in her right pleural cavity (the space between the lungs and the chest wall). She became feverish again, and the blood, urine, and sputum cultures were repeated. A procedure was performed to remove the fluid from her pleural cavity using a needle. The fluid turned out to be purulent, suggesting infection, probably with MRSA. Her condition was called "empyema."

She had a chest tube inserted, and she was told that she would probably need a procedure later to clean her pleural cavity once her condition further improved. Her oxygen needs diminished after the fluid was removed, and her fever reduced.

The blood cultures and urine tests did not show any bacteria. MRSA was again found on the culture of her sputum.

Eventually, she was slowly weaned off mechanical ventilation and then released from the respirator. She still had the chest tube in place, but her breathing was much better, and the pneumonia was clearly improving according to the chest X-rays. She developed back pain at the site of the chest tube insertion. Two days after her release from the respirator she was able to have a clear, liquid diet. The plan was then to perform a procedure called "decortication" to cleanse her pleural cavity of the remaining purulent material once she was sufficiently recovered.

Jeanne was feeling stronger the following week, and her decortication was done by the middle of the week. The procedure went without complications, but she was in severe pain afterward. An IV pump for self-administering morphine was set up, and finally, she had more control over her back pain. She still had a chest tube, and the plan was to remove it once the amount of fluid that was draining diminished.

Five days later, the chest tube was removed, and Jeanne was eating solid food. The analgesic pump was removed, and she was able to sit in her chair and spend a few hours there daily. Since she had improved significantly, she was transferred to another floor. The discharge planner visited her and told her she would go to a rehabilitation center for a while before returning to her assisted-living complex. Her urinary catheter was removed, as she was able to go to the toilet by herself.

The physical therapist worked with her on a daily basis. Initially, Jeanne was not able to walk much. In fact, she was not even able to stand up without having her legs give out. Little by little, she improved and gained the confidence to walk the ward floor several times a day, at first with a walker and then by herself. Her IV antibiotics were continued through a long-term IV line placed for that purpose. She finally was encouraged to go to a rehabilitation center, and she arranged for one of her relatives to visit the facilities prior to choosing one.

The following week, she was transferred to the rehabilitation center, where she spent two weeks finishing her IV antibiotics while making new friends and doing an increased amount of physical therapy. She went to see her primary care physician and then the infectious diseases specialist the next day. The infectious diseases specialist reviewed her chest X-rays and laboratory workup and told her that he was pleased with her improvement. He discontinued her IV antibiotics and removed the IV line that she had been using for antibiotic administration. She was to follow up with him two weeks later for another chest X-ray and more laboratory work, which included a repeat of blood cultures.

Jeanne was told that she could return to her assisted-living community. The next day, she returned to her home. Her neighbors greeted her and gave her a small, welcome-home party. Her laboratory workup and blood cultures were negative. In the following months after her illness, Jeanne did not have a recurrence of her infection.

Jeanne's case illustrates the potential of MRSA to produce severe disease. In the end, Jeanne's case turned out well. However, sometimes complications arise with elderly patients due to the relatively compromised state of their immune systems. Many of those patients have several, pre-existing, chronic or debilitating diseases. Some of them suffer from evident malnutrition due to feeding problems, inability to swallow properly, or simply because they do not have much appetite. Those patients have a poor prognosis if they also have an acute infection of Staph or MRSA.

Chapter X

Why Do We Have More Staph and MRSA Infections?

Despite the large amount of information available on Staph and MRSA, more and more people develop these infections. Why? The answer is probably a mixture of lack of information, prevention, and hygiene.

Most people who have these infections for the first time and are hospitalized tell me they have heard from the news that the infections could happen, but they were not aware that they could spread so quickly and produce so much damage within hours or days; they did not know that washing their hands and having proper hygiene were so important.

The average time it takes for a person to go to the ER seeking attention for a skin infection is between three to seven days. These people also invariably say they thought it would get better. Some others had been in the ER and had their lesions lanced but returned because the lesion got worse despite oral antibiotics. Some others said they waited because they did not have insurance and did not want to pay a big bill. Some try to blame others for their skin infections—their employers, their coworkers, or their doctors. Sometimes it seems like they don't want to be responsible for their own care; they blame the system for their infections and the outcome.

Conclusions

Staph and MRSA infections are a source of significant morbidity and mortality. In certain areas, the number of cases of MRSA skin infections is reaching epidemic proportions. It is a public health threat that must not be ignored because it is potentially lethal. To effectively fight this infection and to reduce its spread requires a significant behavioral change. These "community-associated" MRSA infections, unlike the old Staph infections that caused less invasive infections, spread extremely rapidly and in a matter of hours can produce significant tissue damage leading to necrotizing lesions. Good hygiene and increased public awareness will help to stop the spread of CA-MRSA. The best way to reduce the spread of MRSA, in addition to treating the infections, is to prevent people from getting it.

Photographs

The photographs that follow were taken of patients who gave their full consent before the pictures were obtained. These photographs represent what is frequently seen in hospitals and clinics across the United States as well as in other parts of the world.

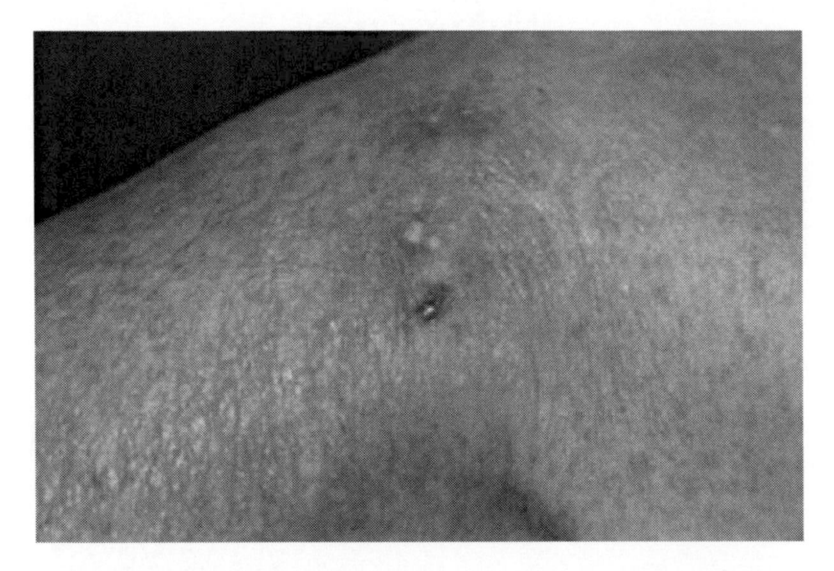

Figure 1. Patient who four days earlier had a "spider bite-like" lesion. Started on antibiotics. Culture revealed Staph.

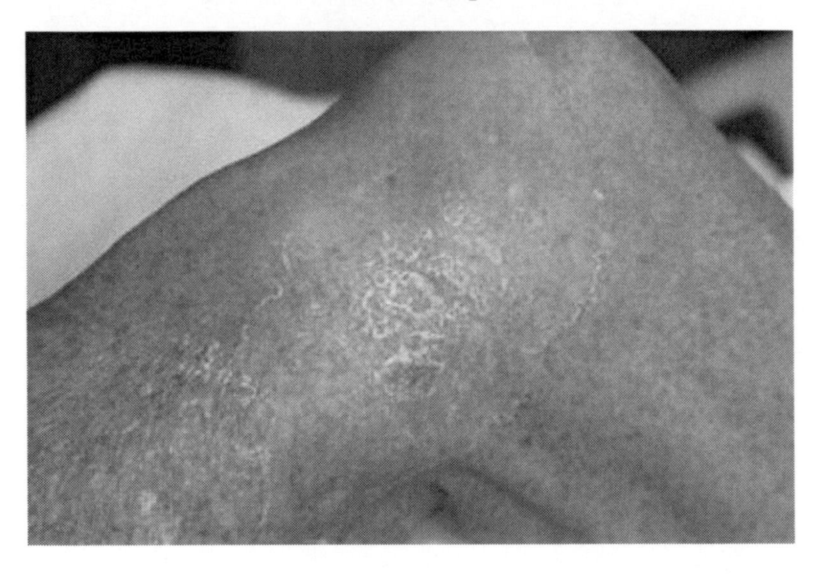

Figure 2. Patient shown in Figure 1, seven days later.

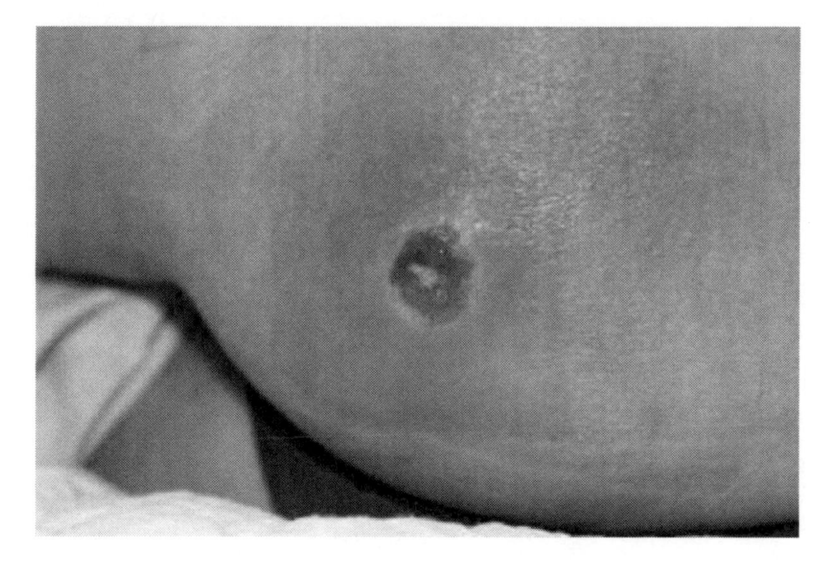

Figure 3. Patient with MRSA buttock ulcer and severe pain four days after a "spider bite-like" lesion.

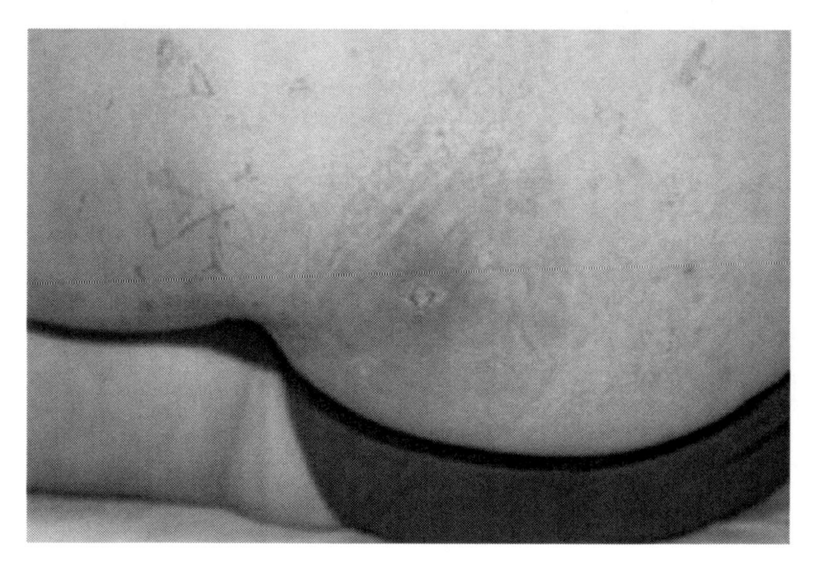

Figure 4. Patient shown in Figure 3, after ten days of antibiotics.

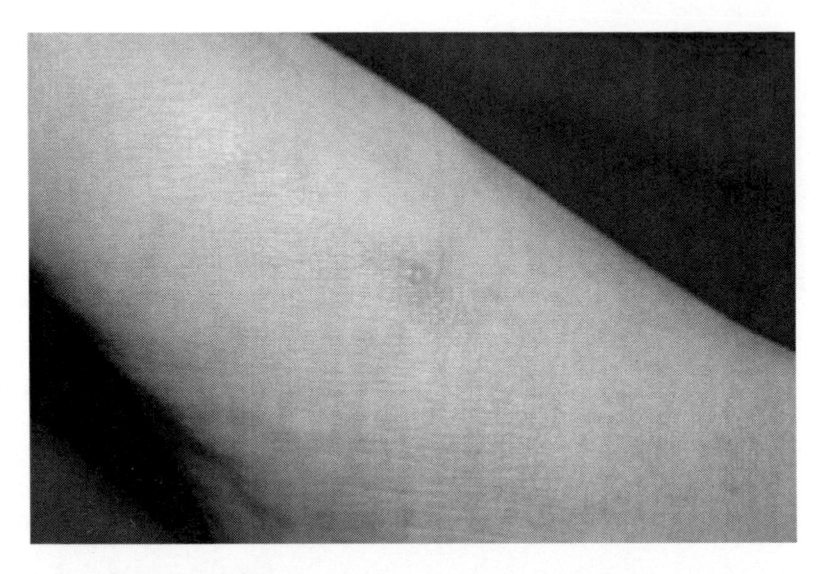

Figure 5. Patient with recurrent "spider bite-like" lesions with itching, redness, and blistering.

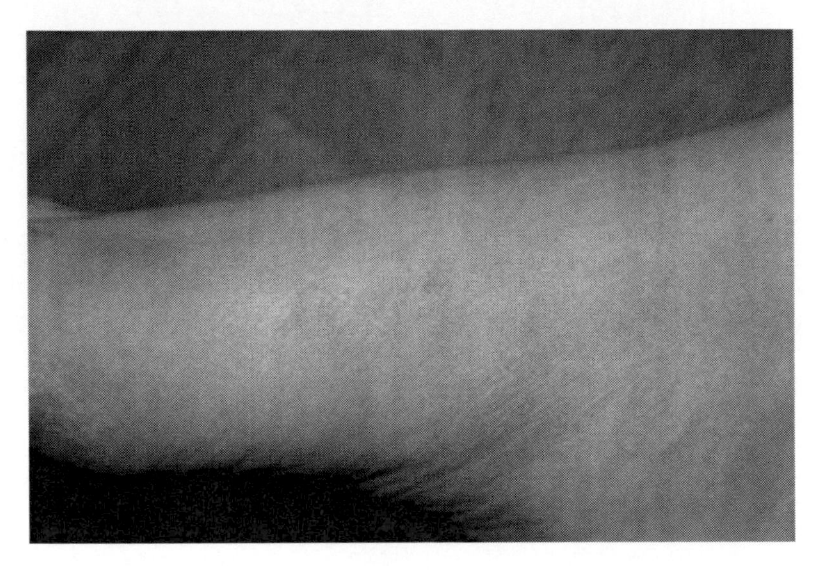

Figure 6. Same lesion shown in Figure 5, twenty-four hours after using mupirocin.

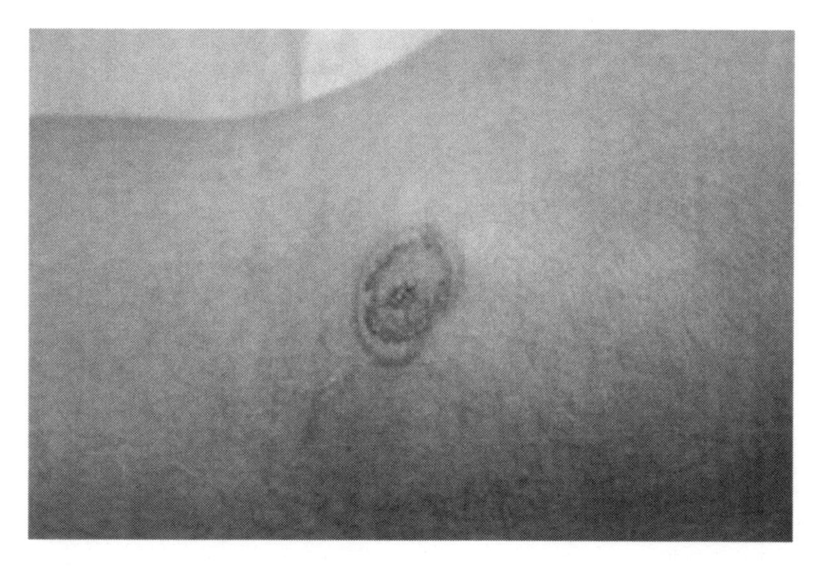

Figure 7. Another patient with a three–week-old lesion. Culture revealed Staph.

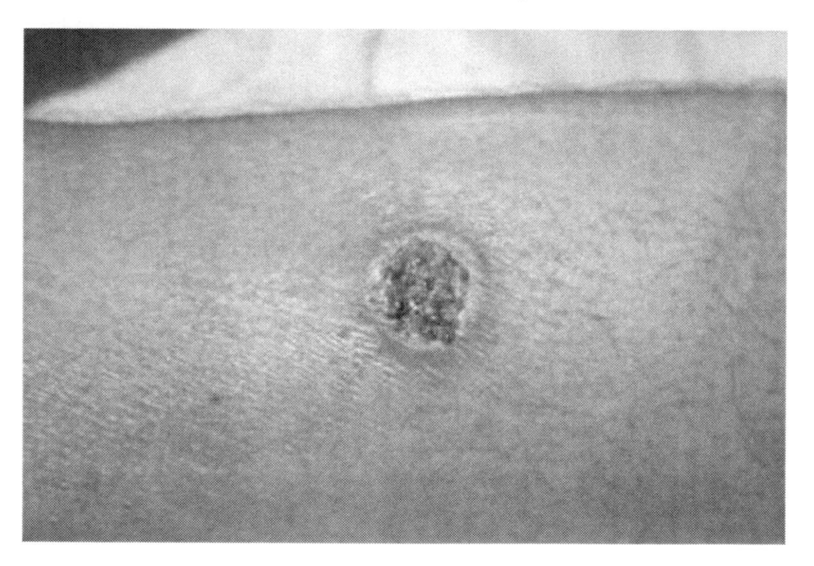

Figure 8. Patient shown in Figure 7, with necrotic lesion.

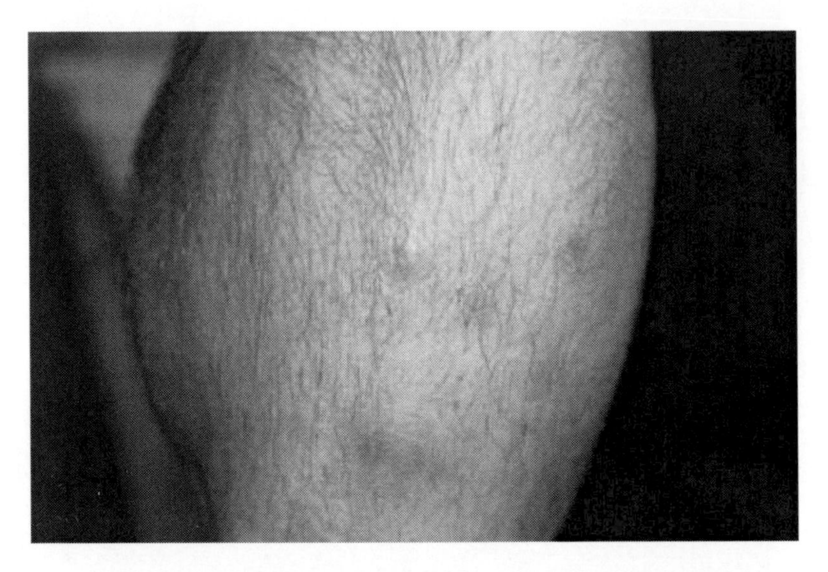

Figure 9. Right forearm of a patient with history of recurrent MRSA infections.

Figure 10. Patient after ten days of blistering and itching. Culture of pustules revealed MRSA

Appendix A

Handwashing

It might seem redundant, but unfortunately most people just do not wash their hands properly. It is preferable to use gloves if you are going to touch a skin lesion or a device that might be contaminated with Staph or MRSA. Do the following before and after touching potentially contaminated surfaces:

If you use soap and water:

1. Apply liquid soap or use a clean bar of soap after putting your hands under running, lukewarm water.
2. Rub your hands for at least fifteen to twenty seconds. Take care to clean the palms, under the fingernails, and the backs of the hands.
3. Rinse your hands well with water.
4. Use a clean paper towel to dry your hands. Use the paper towel to turn off the faucet so as to avoid contaminating your hands again!

If you use a sanitizer, it is preferable to use only the alcohol-based ones.

Appendix B

Intranasal Regimens for Decolonization

Mupirocin

Mupirocin is a topical antibacterial sold at 2 percent concentration in cream and ointment forms. The ointment is better than the cream when used in the nostrils. Before using this medicine you need to contact your healthcare provider, especially if you are pregnant, have multiple allergies, or if you take other medicines.

To apply the ointment:

Wash your hands, blow your nose, and wash your hands again.

Put a small amount (one-fourth of an inch) of mupirocin ointment on your clean, little finger or a cotton bud (if you are applying it to another person use only the cotton bud method, don't use your fingers!), and apply it all around the inside of the nostril. Repeat with the other nostril. This should be done three to four times a day as the ointment melts down. The ointment can cause sneezing or a stinging sensation. If the stinging sensation does not disappear, stop using the ointment and contact your healthcare provider.

Wash your hands after application. This ointment should to be used for approximately five to seven days consecutively, together with some oral antibiotics prescribed by your healthcare provider.

Throw away any ointment left over after finishing the treatment.

Keep the ointment away from children. If you get the ointment in your eyes, rinse with water and call your healthcare provider immediately.

Note that effective decolonization of the nose entails the use of mupirocin and oral antibiotics. Mupirocin alone will probably not be sufficient to successfully eradicate Staph or MRSA from the nostrils.

Bacitracin

Bacitracin has been used but is not very effective against MRSA. Its use for decolonization is not justified.

Appendix C

Topical Regimens for Skin Decolonization

Chlorhexidine Gluconate

Chlorhexidine gluconate is commercially available at 4 percent concentration. In patients that have skin colonization (carriage), chlorhexidine gluconate has been used successfully to eradicate MRSA, although it is less active than 70 percent ethanol and povidone/iodine. Povidone/iodine is not used anymore as its use was shown to lead to systemic absorption and thyroid abnormalities. Showering twice a day with chlorhexidine gluconate is recommended to reduce the skin's bacterial load.

Use a chlorhexidine gluconate 4 percent solution instead of soap during showering. It can also be used as a shampoo, but it is better to avoid using it as a shampoo as it can get in the eyes and ears. First, rinse skin with water. and then apply and wash gently with a small amount of chlorhexidine gluconate solution from the neck down. Rinse well with warm water, and dry the skin. Use it once a day for five to seven days. It can be used intermittently for a period of one week by persons who have constant recurrence of skin infections. Keep it away from the head and face since permanent eye injury may occur after prolonged contact. Also, middle ear contact has led to deafness. Discontinue its use immediately if redness, irritation, or skin allergies appear, and seek medical advice. Hypersensivity is commonly seen in the genital area.

Hexachlorophene

Hexachlorophene is now available only by prescription as its use was associated with neurological deficits in infants in the past. It should not be applied to open skin surfaces.

Retapamulin

Retapamulin is a topical antibiotic from a new class of pleuromutilin antibiotics. It has been approved for the topical treatment of bacterial skin infections, such as impetigo. Impetigo ("school sores") is a superficial, bacterial infection produced by *S. aureus* (MSSA only) and sometimes due to *Streptococcus pyogenes* (Group A). It commonly affects children two to six years old and appears as bullous (blistering) and non-bullous (non-blistering) lesions of the skin. Retapamulin should not be used to treat MRSA because it is not effective against it.

Soaps

Antibacterial soaps contain antimicrobial substances, which are either triclosan or triclocarban. The use of these soaps and soaps in general help in the prevention and transmission of infections in general. However, there is no firm data showing that commercially–available, antibacterial soaps (which contain only a small concentration of antimicrobials) are effective in preventing skin infections.

Bleach

The use of diluted bleach to prevent skin infections is not recommended due to the potential of producing severe skin irritation and damage.

Tea Tree Oil

Tear tree oil is an essential oil from the Australian plant *Melaleuca alternifolia*, which possesses some antimicrobial properties. Its use in the form of soaps and creams has been suggested to prevent boils, acne, and furunculosis. However, some patients have experienced severe allergic reactions to the oil, and its use may not be safe.

Antibiotics Used for Decolonization

A combination of either rifampin and trimethoprim-sulfamethoxazole or rifampin and doxycycline is commonly used. These combinations are inexpensive and should be given together with the intranasal mupirocin. Other antibiotics, such as a quinolone (ciprofloxacin or levofloxacin) or even linezolid, can be used if the MRSA is known to be susceptible. However, these medications are much more expensive, and their use is not routinely recommended as it can contribute to the development of microbial resistance.

Appendix D

Virulence Factors of MSSA and MRSA

Toxins Acting on Cell Membranes

The following all produce cell membrane damage: α-toxin, β-toxin, δ-toxin, and Panton-Valentine leukocidin (see below).

Panton-Valentine Leukocidin (PVL)

Panton-Valentine leukocidin is a cytotoxin (exotoxin) that causes necrotic lesions involving the skin and other tissues. It is harbored mainly by CA-MRSA isolates. The toxin induces pores in the membranes of cells, whose contents then leak and cell death ensues. PVL causes leukocyte destruction and necrotizing pneumonia. It is produced from the genetic material of a bacteriophage after infection of *S. aureus*. Recent studies have challenged the view that PVL is a major factor on *S. aureus* virulence.

Pyrogenic Toxin Superantigens

Pyrogenic toxin superantigens are staphylococcal enterotoxins, which produce food poisoning and toxin TSST-1, which produces Toxic Shock Syndrome.

Exfoliative Toxins

Exfoliative toxins are two epidermolytic exotoxins—A and B. They cause the cleavage of desmoglein-1, which is a protein that holds the granulosum and spinosum layers of the skin together. The result is the detachment of the epidermal layer of the skin with blister formation as in Staphylococcal Scalded-Skin Syndrome.

Staphyloxanthin

Staphyloxanthin is the carotenoid pigment that gives *S. aureus* its golden yellow color. Some studies have shown that staphyloxanthin impairs leukocyte microbicidal activity and promotes virulence through its antioxidant activity. It is also an inhibitor of human squalene synthase (to inhibit cholesterol synthesis) and was shown to reduce staphyloxanthin synthesis *in vitro*. Subsequent studies *in vivo* were done, and some promising results were obtained. More research is needed before this approach can translate into some form of treatment.

Phenol-soluble Modulins

Phenol-soluble modulins are a family of toxins produced by CA-MRSA and thought to be a key virulence for MRSA. Virulent CA-MRSA strains are associated with higher production of the protein toxins. Production of PSM-α-protein was associated with enhanced virulence and destruction of leukocytes.

Microbial Surface Components Recognizing Adhesive Matrix Molecules (MSCRAMMs)

MSCRAMMs attach to collagen, fibrinogen, and fibronectin and mediate adherence to host tissues. They probably play a role in prosthetic device infections and in endocarditis and joint infections.

Other

Other virulence factors of MSSA and MRSA include elastases, proteases, lipases, peptidoglycan, lipoteichoic acid, phospholipase C, metalloproteases, hyaluronate lyase, and several others.

Appendix E

Microbiological and Epidemiological Aspects of Staphylococcal Resistance

Based on minimal inhibitory concentrations (MIC), bacteria are classified as susceptible, intermediate, or resistant. The breakpoint helps in the decision of what to use when treating a particular infection. *Susceptible* bacteria are probably going to respond to the antibiotic used. *Intermediate* bacteria may or may not respond to the antibiotic used. *Resistant* bacteria are likely to fail to respond to the achievable blood drug concentration of the antibiotic. The breakpoint is set by the FDA and certified by the Clinical Laboratory Standards Institute.

S. aureus is isolated from a clinical sample (of wound material, pus, sputum, etc.) by plating the material on blood agar and mannitol salt agar (contains 7–9 percent NaCl). The *S. aureus* colonies appear yellow on the mannitol salt agar. Then other confirmatory tests are used: catalase (all Staphylococci are positive), coagulase (fibrin clot formation), phosphatase, DNase, and lipase. The catalase test means that the bacterium is able to calalyze the reaction from hydrogen peroxide to oxygen and water. The test is useful to distinguish staphylococci from streptococci and enterococci, as they don't produce catalase.

Most *S. aureus* are coagulase positive, which helps to differentiate them from other less pathogenic staphylococci (coagulase-negative staphylococci).

For identification of nasal colonization, a selective media for MRSA, such as Chromogenic Agar, can be used, and no further tests are required. Rapid tests based on polymerase chain reaction techniques (DNA amplification) can be used, but they are more expensive.

Resistance to the antibiotic clindamycin is sometimes inducible in *S. aureus*. This can be demonstrated in vitro after exposure to a macrolide antibiotic. For this the D-zone test is used. In this test, erythromycin and clindamycin disk pairs are placed in an agar plate that is seeded with *S. aureus* beforehand. The bacterial growth is measured, and if there is a flattening of the growth zone of inhibition near the clindamycin disk, the test indicates inducible resistance (zone is "D" shaped, hence D-zone test). This strain would otherwise be reported as susceptible using conventional susceptibility testing.

HA-MRSA appeared in the 1970s, and CA-MRSA appeared in the 1990s. There are several differences between these two variants, although the differences are beginning to blur. CA-MRSA is mostly clonal (one clone predominated), is sensitive to many antibiotics, and has the Panton-Valentine leukocidin gene. HA-MRSA is polyclonal and resistant to many antibiotics.

In CA-MRSA strains, the methicillin resistance gene (*mec*) is carried in the staphylococcal cassette element (SCC) type IVa. In HA-MRSA, the *mec* gene is carried in SCC type I, II, and III.

The USA300 strain of MRSA, which has the SCC type IVa and the PVL gene, appears to be causing the majority of the CA-MRSA cases in the United States. Another virulent strain of MRSA is USA400, which has the PVL gene and produces other toxins, including enterotoxin H, which can cause Toxic Shock Syndrome.

Appendix F

Molecular Aspects of Staphylococcal Resistance

Resistance to Penicillin

S. aureus resistance to penicillin is mediated by penicillinase (a type of β-lactamase) production. The enzyme destroys the β-lactam ring of the antibiotic molecule. Penicillins are no longer used to treat *S. aureus* infections even if the *in vitro* test shows susceptibility. This is due to the fact that they may be able to produce penicillinase during treatment, leading to a treatment failure.

Resistance to Methicillin

S. aureus resistance to methicillin (thus MRSA) is mediated by an alteration in the penicillin binding protein 2 (PBP2). This site (PBP2A or PBP2') is encoded by the gene *mecA* found in a mobile genetic element known as the SCC *mec*. This altered PBP2 has a lower affinity for binding β-lactam antibiotics—such as methicillin, oxacillin, nafcillin, dicloxacillin, cephalotin, and cefazolin—leading to resistance.

Resistance to Vancomycin

Resistance to vancomycin is rare. *S. aureus* resistance to vancomycin is mediated through the *vanA* gene that was initially described in enterococci. The *vanA* gene encodes for an enzyme that produces an alternative peptidoglycan to which vancomycin will not bind. These strains of vancomycin-resistant *S. aureus* (VRSA) are not affected by vancomycin, so alternative antibiotics need to be used. At present, only five cases have been seen in the United States.

Vancomycin-intermediate *S. aureus* (VISA, also called GISA or glycopeptide-intermediate *S. aureus*) was initially described in Japan in 1996 (Hiramatsu K, et al. 1997. *J. Antimicrob. Chemother.* 40:135-6).. Resistance in VISA is not mediated by a specific gene but is due to several mutations that lead to changes in cell-wall metabolism, which serve as dead-end binding sites for vancomycin. The antibiotic, therefore, cannot reach its target and is rendered ineffective.

Further Reading

Adem PV, Montgomery CP, Husain AN, Koogler TK, Arangelovich V, Humilier M, Boyle-Vavra S, and Daum RS. 2005. *Staphylococcus aureus* sepsis and the Waterhouse-Friderichsen syndrome in children. *N. Engl. J. Med.* 353:1245–51.

Benner EJ, and Kayser FH. 1968. Growing clinical significance of methicillin-resistant *Staphylococcus aureus. Lancet* 2:741–44.

Bischoff WE, Wallis ML, Tucker BK, Reboussin BA, Pfaller MA, Hayden FG, and Sherertz RJ. 2006. "Gesundheit!" Sneezing, common colds, allergies, and *Staphylococcus aureus* dispersion. *J. Infect. Dis.* 194:1119–26.

Bootsma MJ, Diekmann O, and Bonten MJ. 2006. Controlling methicillin-resistant *Staphylococcus aureus*: Quantifying the effects of interventions and rapid diagnostic testing. *Proc. Natl. Acad. Sci. USA* 103:5620–25.

Boyce JM. 2001. MRSA patients: proven methods to treat colonization and infection. *J. Hosp. Infect.* 48 (Suppl. A):S9–S14.

Chang HR. 2004. Cellulitis (Letter). *N. Engl. J. Med.* 350:2423.

Chang HR, and Bistrian BR. 1998. The role of cytokines in the catabolic consequences of infection and injury. *J. Parenteral Enteral Nutr.* 22:156-66.

Chang S, Sievert DM, Hageman JC, Boulton ML, Tenover FC, Downes FP, Shah S, Rudrik JT, Pupp GR, Brown WJ, Cardo D, Fridkin SK, and the Vancomycin-resistant *Staphylococcus aureus* Investigative Team. 2003.

Infection with vancomycin-resistant *Staphylococcus aureus* containing the *vanA* resistance gene. *N. Engl. J. Med.* 348:1342–47.

Coia JE, Duckworth GJ, Edwards DI, Farrington M, Fry C, Humphreys H, Mallagan C, Tucker DR, and Joint Working Party of the British Society of Antimicrobial Chemotherapy, Hospital Infection Society, Infection Control Nurses Association. 2006. Guidelines for the control and prevention of methicillin-resistant *Staphylococcus aureus* (MRSA) in healthcare facilities. *J. Hosp. Infect.* 63 (Suppl. 1):S1–S44.

Datta R, and Huang SS. 2008. Risk of infection and death due to methicillin-resistant *Staphylococcus aureus* in long-term carriers. *Clin. Infect. Dis.* 47:176–81.

Eriksen NH, Espersen F, Rosdahl VT, and Jensen K. 1995. Carriage of *Staphylococcus aureus* among 104 healthy persons during a 19-month period. *Epidemiol. Infect.* 115:51–60.

Francis JS, Doherty MS, Lopatin U, Johnston CP, Sinha G, Ross T, Cai M, Hansel NN, Perl T, Ticehurst JR, Carroll K, Thomas DL, Nuermberger E, and Bartlett JG. 2005. Severe community-onset pneumonia in healthy adults caused by methicillin-resistant *Staphylococcus aureus* carrying the Panton-Valentine leukocidin genes. *Clin. Infect. Dis.* 40:100–107.

Gillet Y, Issartel B, Vanhems P, Fournet JC, Lina G, Bes M, Vandenesch F, Piemont Y, Brousse N, Floret D, and Etienne J. 2002. Association between *Staphylococcus aureus* strains carrying gene for Panton-Valentine leukocidin and highly lethal necrotizing pneumonia in young immunocompetent patients. *Lancet* 359:753–59.

Graham PL, Lin SX, and Larson EL. 2006. US population-based survey of *Staphylococcus aureus* colonization. *Ann. Intern Med.* 144:318–25.

Grundmann H, Aires-de-Souza M, Boyce J, and Tiemersma E. 2006. Emergence and resurgence of methicillin-resistant *Staphyloccoccus aureus* as a public-health threat. *Lancet* 368:874–85.

Hardy KJ, Oppenheim BA, Gossain S, Gao F, and Hawkey PM. 2006. A study of the relationship between environmental contamination with methicillin-resistant *Staphylococcus aureus* (MRSA) and patients' acquisition of MRSA. *Infect. Control Hosp. Epidemiol.* 27:127–32.

Henderson DK. 2006. Managing methicillin-resistant staphylococci: a paradigm for preventing nosocomial transmission of resistant organisms. *Am. J. Med.* 119 (6 Suppl. 1):S45–S52.

Hiramatsu K, Hanaki H, Ino T, Yabuta K, Oguri T, and Tenover FC. 1997. Methicillin-resistant *Staphylococcus aureus* clinical strain with reduced vancomycin susceptibility. *J. Antimicrob. Chemother.* 40:135-6.

Huang R, Mehta S, Weed D, and Price CS. 2006. Methicillin-resistant *Staphylococcus aureus* survival on hospital fomites. *Infect. Control Hosp. Epidemiol.* 27:1267–69.

Huskins WC, and Goldmann DA. 2005. Controlling methicillin-resistant Staphylococcus aureus, aka "Superbug." *Lancet* 365:273–75.

Huws SA, Smith AW, Enright MC, Wood PJ, and Brown MR. 2006. Amoebae promote persistence of epidemic strains of MRSA. *Environ. Microbiol.* 8:1130–33.

Kaneko J, and Kamio Y. 2004. Bacterial two-component and hetero-heptameric pore-forming cytolytic toxins:

structures, pore forming mechanism, and organization of the genes. *Biosci. Biotechnol. Biochem.* 68:981–1003.

Kazakova SV, Hageman JC, Matava M, Srinivasan A, Phelan L, Garfinkel B, Boo T, McAllister S, Anderson J, Jensen B, Dodson D, Lonsway D, McDougal LK, Arduino M, Fraser VJ, Killgore G, Tenover FC, Cody S, and Jernigan DB. 2005. A clone of methicillin-resistant *Staphylococcus aureus* among professional football players. *N. Engl. J. Med.* 352:468–75.

Kitai S, Shimizu A, Kawano J, Sato E, Nakano C, and Kitagawa H. 2005. Prevalence and characterization of *Staphylococcus aureus* and enterotoxigenic *Staphylococcus aureus* in retail raw chicken meat throughout Japan. *J. Vet. Med Sci.* 67:269–74.

Kluytmans J, van Blekum A, and Verbrugh H. 1997. Nasal carriage of *Staphylococcus aureus*: epidemiology, underlying mechanisms, and associated risks. *Clin. Microbiol. Rev.* 10:505–20.

Kuehnert MJ, Kruszon-Moran D, Hill HA, McQuillan G, McAllister SK, Fosheim G, McDougal LK, Chaitram J, Jensen B, Fridkin SK,

Killgore G, and Tenover FC. 2006. Prevalence of *Staphylococcus aureus* nasal colonization in the United States, 2001–2002. *J. Infect. Dis.* 193:172–79.

Lina G, Piemont Y, Godail-Gamot F, Bes M, Peter MO, Gauduchon V, Vandenesch F, and Etienne J. 1999. Involvement of Panton-Valentine leukocidin-producing *Staphylococcus aureus* in primary skin infections and pneumonia. *Clin. Infect. Dis.* 29:1129–32.

Liu CI, Liu GY, Song Y, Yin F, Hensler ME, Jeng WY, Nizet V, Wang AH, and Oldfield E. 2008. A cholesterol biosynthesis inhibitor blocks *Staphylococcus aureus* virulence. *Science* 309: 1391–94.

Lowy FD. 1998. *Staphylococcus aureus* infections. *N. Engl. J. Med.* 339:520–32.

Miller LG, Perdreau-Remington F, Rieg G, Mehdi S, Perlroth J, Bayer AS, Tang AW, Phung TO, and Spellberg B. 2005. Necrotizing fasciitis caused by community-associated methicillin-resistant *Staphylococcus aureus* in Los Angeles. *N. Engl. J. Med.* 352:1445–53.

Miller MA, Dascal A, Portnoy J, and Mendelson J. 1996. Development of mupirocin resistance among methicillin-resistant *Staphylococccus aureus* after widespread use of nasal mupirocin ointment. *Infect. Control Hosp. Epidemiol.* 17:811–13.

Moran GJ, Krishnadasan A, Gorwitz RJ, Fosheim GE, McDougal LD, Carey RB, Talan DA, and for the EMERGEncy ID NET Study Group. 2006. Methicillin-resistant *S. aureus* infections among patients in the Emergency Department. *N. Engl. J. Med.* 355:666–74.

Muto CA, Jernigan JA, Ostrowsky BE, Richet HM, Jarvis WR, Boyce JM, and Farr BM. 2003. SHEA guideline for preventing nosocomial transmission of multidrug-resistant strains of *Staphylococcus aureus* and enterococcus. *Infect. Control Hosp. Epidemiol.* 24:362–86.

Rihn JA, Michaels MG, and Harner CD. 2005. Community-acquired methicillin-resistant *Staphylococcus aureus*: an emergent problem in the athletic population. *Am. J. Sports Med.* 33:1924–29.

Sievert DM, Rudrik JT, Patel JB, McDonald LC, Wilkins MJ, and Hageman JC. 2008. Vancomycin-resistant *Staphylococcus aureus* in the United States, 2002–2006. *Clin. Infect. Dis.* 46:675–77.

Tiemersma EW, Bronzwaer SL, Lyytikainen O, Degener JE, Schrijnemakers P, Bruinsma N, Monen J, Witte W, Grundman H, and European Antimicrobial Resistance

Surveillance System Participants. 2004. Methicillin-resistant *Staphylococcus aureus* in Europe, 1999–2002. *Emerg. Infect. Dis.* 10:1627–34.

US Centers for Disease Control and Prevention. 2003. Methicillin-resistant *Staphylococcus aureus* infections among competitive sports participants—Colorado, Indiana, Pennsylvania, and Los Angeles County, 2000–2003. *MMWR Morb. Mortal. Wkly. Rep.* 52:793–95.

US Centers for Disease Control and Prevention. 2003. Methicillin-resistant *Staphylococcus aureus* infections in correctional facilities—Georgia, California and Texas, 2001–2003. *MMWR Morb. Mortal. Wkly. Rep.* 52:992–96.

Vandenesch F, Naimi T, Enright MC, Lina G, Nimmo GR, Heffernan H, Liassine N, Bes M, Greenland T, Reverdy ME, and Etienne J. 2003. Community-acquired methicillin-resistant *Staphylococcus aureus* carrying Panton-Valentine leukocidin genes: worldwide emergence. *Emerg. Infect. Dis.* 9:978–84.

Van Loo IHM, Diederen BMW, Savelkoul PHM, Woudenberg JHC, Roosendaal R, van Belkum A, van Keulen PHJ, and Kluytmans JAJW. 2007. Methicillin-resistant *Staphylococcus aureus* in meat products, the Netherlands. *Emerg. Infect. Dis.* 13:1753–55.

Von Eiff C, Becker K, Machka K, Stammer H, and Peters G. 2001. Nasal carriage as a source of *Staphylococcus aureus* bacteremia. Study Group. *N. Engl. J. Med.* 344:11–16.

Wang R, Braughton KR, Kretschmer D, Bach TH, Queck SY, Li M, Kennedy AD, Dorward DW, Klebanoff SJ, Peschel A, DeLeo FR, and Otto M. 2007. Identification of novel cytolytic peptides as key virulence determinants of community-acquired MRSA. *Nature Medicine* 13:1510–14.

Weigelt J, Kaafarani HMA, Itani KFM, and Swanson RN. 2004. Linezolid eradicates MRSA better than vancomycin from surgical-site infections. *Am. J. Surg.* 188:760–66.

Wunderink RG, Rello J, Cammarata SK, Croos-Cabrera RV, and Kollef MH. 2003. Linezolid vs vancomycin: Analysis of two double-blind studies of patients with methicillin-resistant *Staphylococcus aureus* nosocomial pneumonia. *Chest* 124:1789–97.

Yu VL, Goetz A, Wagener M, Smith PB, Rihs JD, Hanchett J, and Zuravleff JJ. 1986. *Staphylococcus aureus* carriage and infection in patients on hemodialysis. *N. Engl. J. Med.* 315:91–96.

Lightning Source UK Ltd.
Milton Keynes UK
172951UK00002B/172/P